# Five Tough and Tiny Seeds

By Steve Metzger

Illustrated by Mike Byrne

**Scholastic Inc.**

Five tough and tiny seeds sat on a slender reed,
Waiting for springtime winds to blow—

WHOOSH! WHOOSH!

Four tough and tiny seeds sat on a slender reed,
Waiting for springtime winds to blow—

WHOOSH! WHOOSH!

One landed near a lake, beside a wriggly snake,

Then there were three tough, tiny seeds.

Three tough and tiny seeds sat on a slender reed,

Waiting for springtime winds to blow—

WHOOSH! WHOOSH!

One landed near a tree,

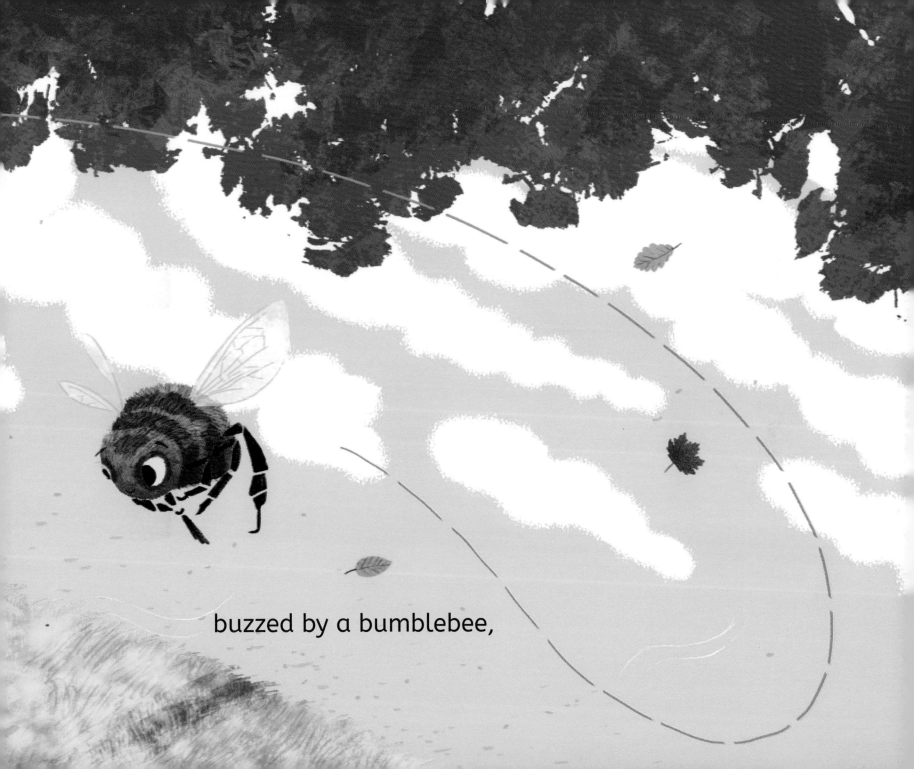

Then there were two tough, tiny seeds.

Two tough and tiny seeds sat on a slender reed,
Waiting for springtime winds to blow—

WHOOSH!
WHOOSH!

One landed by a road, next to a slimy toad,

Then there was one tough, tiny seed.

One tough and tiny seed sat on a slender reed,
Waiting for springtime winds to blow—

WHOOSH! WHOOSH!

Then there were no tough, tiny seeds.

Five tough and tiny seeds blown from a slender reed,

Raindrops are falling from the sky—

DRIP!
DROP!

With bright and warmer days, now comes the month of May,

Now it's the time for seeds to grow.

Five seeds begin to sprout, just as the sun comes out,

What will they all turn out to be?

Growing taller by the hour,

each one a different flower,

rose

dandelion

sunflower

cosmos

wild carrot

That's what became of five tough, tiny seeds.

# Life Cycle: From Seeds to Plants

 **1 Seeds**

Seeds come in different sizes and shapes. Inside every seed is a tiny plant just waiting to grow. Some seeds become flowers, such as daffodils, roses, sunflowers, and tulips.

 **2 Seeds on the Move**

Some seeds fall close to the plants that made them. Other seeds travel to distant locations. This can happen when seeds fall into rivers or streams, or when animals carry them to new places.

 **3 Germination**

Seeds must be in the soil—or resting on top of the soil—in order to sprout. When it rains, the coats on the outsides of the seeds are softened. Afterward, the sun helps warm the ground. Then the seed coats open, and the seeds begin to grow.

 **4 Roots and Shoots**

Roots grow down into the ground and absorb water and minerals from the soil. Shoots are the new parts of plants that come up from the ground.

##  5  Leaves

Green leaves unfold from the plants' stems and grow up toward the sun. Leaves take in sunlight and produce food for the plant through a process called *photosynthesis*.

##  6  Flowers

When flowering plants are fully grown, their buds blossom into colorful flowers, where new seeds grow.

##  7  Pollination

Before seeds can grow, they must be *pollinated*. Sometimes, the wind blows pollen from flower to flower. Other times, the colors and scents of flowers attract insect pollinators, such as bees. When bees visit a flower, pollen lands on their bodies. Afterward, they carry the pollen to other flowers when visiting them for nectar.

##  8  Seeds Get Bigger

Seeds grow inside the flowers of plants. As the seeds get bigger, fruit or pods grow around them. When the fruit or pods break open, the seeds are ready to become new plants. The seed-to-plant cycle begins again!

To Diane Bruce; and John, Alyssa, and Sam Sinclair. — S. M.

For Oscar and Harry. — M. B.

Text copyright © 2018 by Steve Metzger

Illustrations copyright © 2018 by Mike Byrne

All rights reserved. Published by Scholastic Inc., *Publishers since 1920*.
SCHOLASTIC and associated logos are trademarks and/or registered trademarks of Scholastic Inc.

The publisher does not have any control over and does not assume any responsibility for author or third-party websites or their content.

No part of this publication may be reproduced, stored in a retrieval system, or transmitted in any form or by any means, electronic, mechanical, photocopying, recording, or otherwise, without written permission of the publisher. For information regarding permission, write to Scholastic Inc., Attention: Permissions Department, 557 Broadway, New York, NY 10012.

ISBN 978-1-338-23099-4

10 9 8 7 6 5 4 3     18 19 20 21 22

Printed in the U.S.A.     40

First printing 2018

Book design by Jennifer Rinaldi